The Adventures of
BIPLOB THE BUMBLEBEE
VOLUME 2

Biplob World Pvt. Ltd.

Publisher Ritika Talwar
Published by Biplob World Pvt. ltd.
© 2024 Biplob World Pvt. Ltd. All Rights Reserved.
Typeset in Sabon LT Std Roman

*Winner of NMCBI Best Children's Author 2021

CONTENTS

Story 1: Power of the Sun
Solar Power

Story 2: Farm to Plate
How food is Grown

Story 3: Reap as you Sow
Why littering is Bad

1
Power
of the
Sun

Farmer Balram was a very worried man. It had not rained for many days and his crops were beginning to dry without water.

There was so little water that even the plants in his garden were suffering now. All the flowers were thirsty but there was nothing they could do.

They knew they would get no water unless it rained.

"I wish we had been more careful in using the water when our well was full. At least we would have had some water to use today," farmer Balram said aloud.

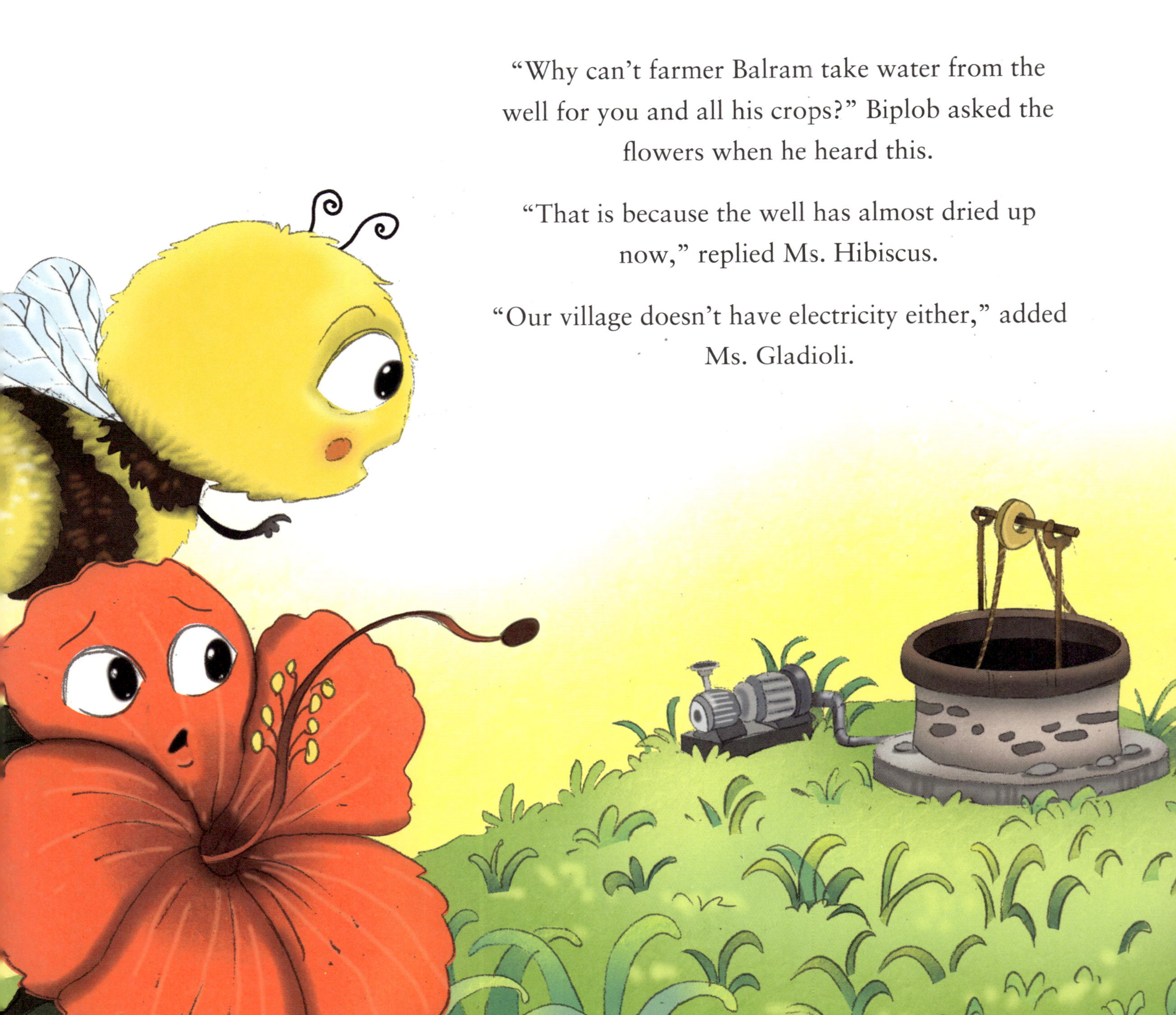

"Why can't farmer Balram take water from the well for you and all his crops?" Biplob asked the flowers when he heard this.

"That is because the well has almost dried up now," replied Ms. Hibiscus.

"Our village doesn't have electricity either," added Ms. Gladioli.

"Yes, so farmer Balram cannot use the pump to draw water from the well," explained Ms. Hibiscus.

"Oh, I'm so thirsty. I think I will faint!" remarked Ms. Sunflower.
"The sun is so bright and hot, even my eyes hurt now," said Ms. Tulip.

Biplob buzzed away thinking, 'How can I help my friends get water?'

As Biplob looked up, he had an idea, "Why don't we use the great power of the sun to get some water?" he asked farmer Balram.

"Don't be silly, Biplob," replied the farmer, "I agree the sun is very powerful. But how can anyone use that energy to make water?"

"Not to make water but generate electricity. We can use the sun's energy, SOLAR POWER, to run the pump!" replied Biplob, "Using solar power the pump can draw water from deep within the well!" he explained.

Farmer Balram thought about it and realised that this might actually be a very good idea.

Farmer Balram promptly went to the market and got a solar panel, some wires and a battery. He then set about connecting these to his water pump.

Biplob buzzed excitedly around him.

"Hush Biplob. I'm unable to concentrate with your buzzing. Off with you! Let me work in peace," said farmer Balram.

Soon the solar water pump was ready.

Farmer Balram said a prayer and switched it on. The pump came on with a flourish and in no time water started gushing out from the pipes into the garden!

Farmer Balram started dancing around the flowers in glee.
"It works, it works!" he rejoiced.

All the plants quenched their thirst and heaved a sigh of relief.

"Thank you Biplob. You came up with such a simple and eco – friendly solution," they smiled.

"Yes, now I can save my crops and never worry about water again," said farmer Balram.

Biplob was happy to have helped his friends yet again.

2
Farm to Plate

'The summer seems to be getting hotter with every passing year,' thought Biplob as he tiredly flew around the garden one hot afternoon.

"Looks like you aren't going to collect your full share of nectar today, Biplob," remarked Ms. Gladioli, looking at him.

"Oh no, Ms. Gladioli," remarked Biplob, "It's just that this heat has sapped all my strength. I don't feel like doing anything at all!"

Before Ms. Gladioli could reply, they heard the voices of little children chattering.

They saw a little girl and boy walk into the garden.

"If we were home, at least we could've found something interesting to put in our project report," said the girl.

"You're right," replied the boy, "Everything is so plain and boring over here." They slowed to a tired shuffle, their energy sapped by the heat!

"The cheek! If it wasn't so hot," piped up Ms. Violet, "I would've shown these outsiders a thing or two!"

"Would you now?" Ms. Hibiscus asked sweetly, "I thought the term 'shrinking violet' was coined seeing how you run away from any situation!"

"Everything is my fault," wailed Ms. Violet hearing this. "Oh, if only I could get away from all these insensitive flowers."

"Now ladies, let's not argue," said Biplob, calming them down. "They're just little children probably visiting the countryside for the first time and don't know much about us or our way of life. Let me talk to them."

Saying this, he zoomed across the garden towards the children, forgetting all about the heat and how tired he was!

"Hello, hello, hello and welcome!" greeted Biplob, doing a curtsy in the air, "I am Biplob, and these are my friends Ms. Violet, Ms. Hibiscus and all the other flowers in this wonderful garden."

"Well, he is quite the showman," muttered Ms. Daisy to the other flowers, who nodded in agreement, looking at Biplob's act.

"Wow! A talking & dancing bee!" exclaimed the girl.

"And he knows the names of these flowers too," added the boy, "Hi Biplob, I'm Aditya and this is my sister, Avantika."

"We are here on a holiday," added Avantika.

"Wow, that's exciting. I've never been on a holiday," replied Biplob, "Now what is this report you were talking about?"

"We have to submit a school report on what we did during our holidays," answered Aditya.

"But this place is too boring to make a good report," added Avantika sadly.

"Oh you couldn't have picked a better place to make a splendid report. Come, I'll show you all that is wonderful in the countryside," buzzed Biplob, the children skipping along beside him.

"Look here. This is our well from where we draw water for the plants. And do you know, the pump works on solar power," said Biplob, as he narrated how farmer Balram had fixed the solar pump to solve their water problem.

The next stop was the wheat and maize fields. "Do you like pizza?" asked Biplob.
"We LOVE pizza," replied the children together.

"When this wheat is ripe, farmer Balram will send it to the mill to be made into flour,"
explained Biplob. "Flour is used to make the delicious chapattis and
yummy pizzas you love!"

"And look there," said Biplob, pointing to the sunflowers.
"They look so beautiful," remarked Aditya.

"Yes, don't they?" continued Biplob. "Did you know the seeds of those flowers are crushed to get a very healthy cooking oil!"

"Sunflower oil!" guessed Avantika, "I've seen pictures of sunflowers on the bottle of oil that mummy cooks with at home."

"This is amazing," said Aditya, "Now we know how everything needed to make a pizza is grown."

"I bet everyone at school will be amazed when we put all this in our report," added Avantika.

"Yes, we have everything needed to make the best summer project report ever," agreed Aditya, jumping with excitement.

"Thank you Biplob," said the smiling children together.

"You're welcome," beamed Biplob, "Please drop by whenever you wish to play in our garden!"

"Yes, we will," the two promised, happily skipping back home.

As for a happy Biplob, he buzzed back to work, the heat all forgotten!

3
Reap as you Sow!

"WHO MADE THIS MESS?" shouted farmer Balram angril

The garden was littered with empty packets of chip
and a broken soft-drink bottle glinting in the sunligh

"A group of strangers ha
a picnic here while yo
were away," answere
Ms. Violet sadly, "This
how they left everythin
once they were done.

"Such a shame!" cried Biplob. "Do people keep their homes this filthy?"

"No, they don't!" replied farmer Balram grimly. "People clean their own homes but throw garbage in public places like roads and gardens!"

Biplob and farmer Balram spent the rest of the afternoon silently cleaning up the garden. No one felt like talking.

Early next morning, farmer Balram left for the fields and Biplob went to meet a friend.

As farmer Balram returned for lunch, he bumped into Biplob. "Looks like someone had a great time!" he said to a happy – looking bee.

Suddenly, they heard someone scream from the garden – they rushed to see who it was.

A teenaged girl was crying, as her foot bled profusely. Her friends all stood around her.

Biplob buzzed to the injured girl, laughing, "Aha! You reap as you sow!"

"She's injured and you're making fun of her?" asked a boy in anger.

"Let's complain to the police that she got injured in your garden!" added another girl haughtily, "Then we'll see who laughs!"

"I'm sorry you got hurt in our garden. Please let me check the injury," farmer Balram pacified them. "Don't be sorry farmer Balram!" Biplob butted in, still laughing.
Before farmer Balram could reply, Ms. Daisy spoke up calmly, "Wait farmer Balram. Look around once!" The entire garden was littered with garbage and empty glass bottles, exactly like the previous day!

"Were you here yesterday too?" asked farmer Balram. "Why? Is there a law against it?" replied the haughty girl.

"No, in fact I'm happy when people visit my garden to enjoy nature," replied farmer Balram, a little angry. "But there surely is a law against LITTERING! Yesterday, you shamelessly littered all over the garden. Today you've done the same thing!"

"Exactly! I'm sure the shard of glass stuck in her foot fits here perfectly!" said Biplob, buzzing towards a broken bottle. It had a piece of glass missing.

"Biplob was right . . . as you sow, so you reap!" repeated farmer Balram, pulling the glass shard out of the girl's foot. "Your habit of littering has injured one of your own friends."

The teenagers looked around the garden and hung their heads in shame. "We are very sorry, sir. We didn't realise the consequences of our actions," the haughty girl said.

"Yes, we promise to clean up everything and never litter ever again," added another boy, his friends nodding solemnly.

Soon enough, the group of friends cleaned up the garden with
farmer Balram and Biplob's help.

"Thank you, Biplob and farmer Balram. I've learnt a valuable lesson today," said the injured girl,
limping on her bandaged foot, as the group got ready to leave.
"We all have!" chimed in her friends together.

The Adventures of Biplob the Bumblebee : Book Series

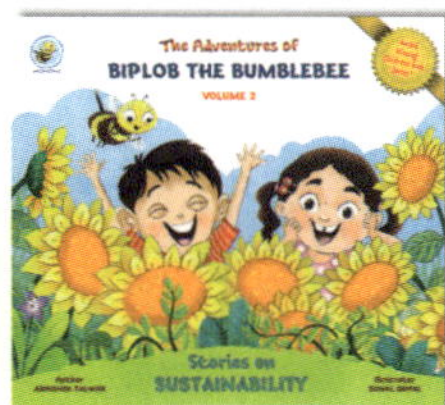
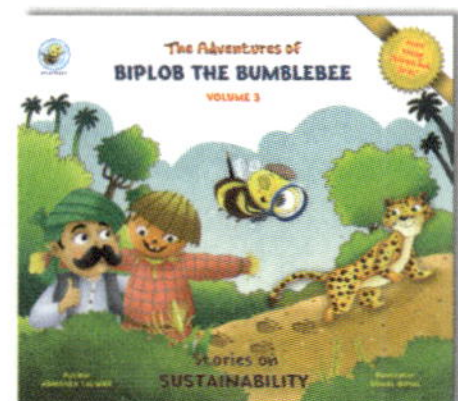
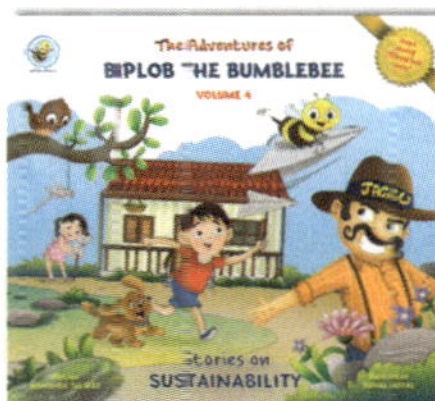

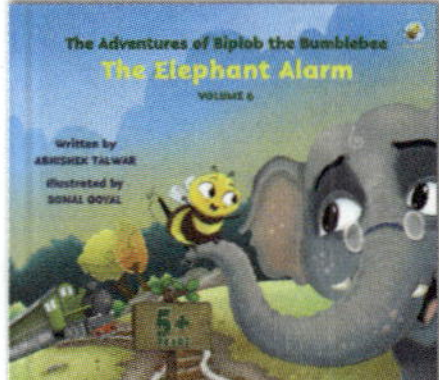

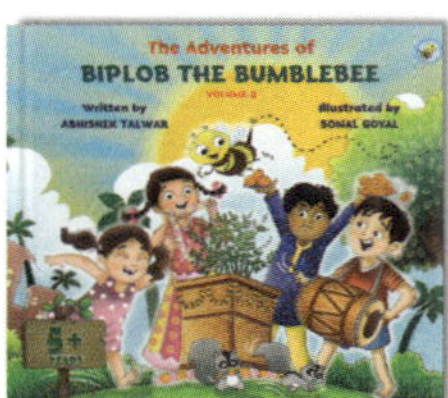
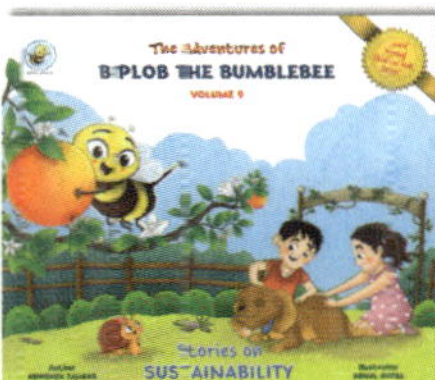

Biplob the Bumblebee : Early Learner Series

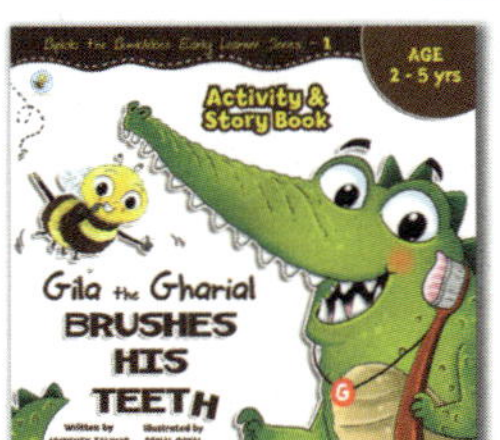

Detective Col. Zoro Series

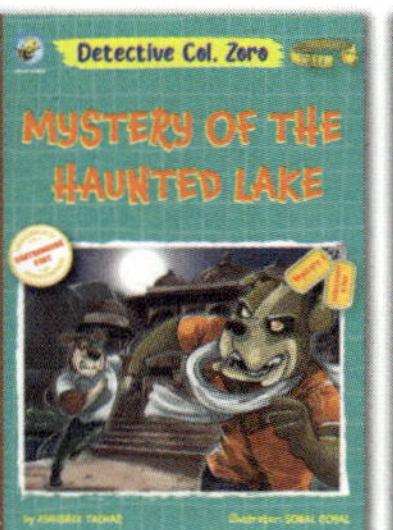

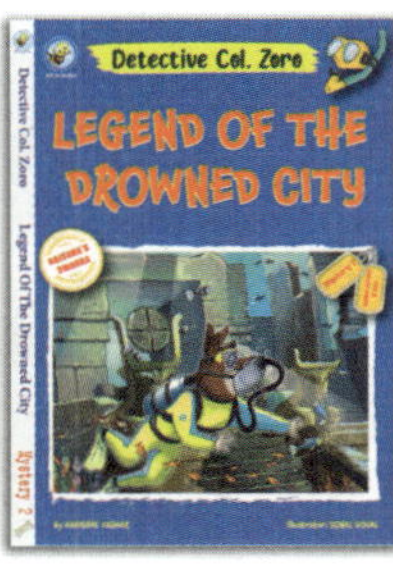

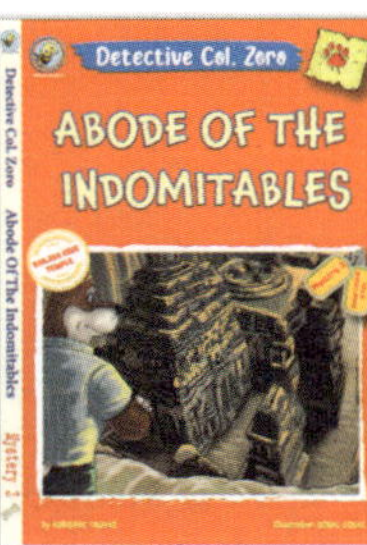

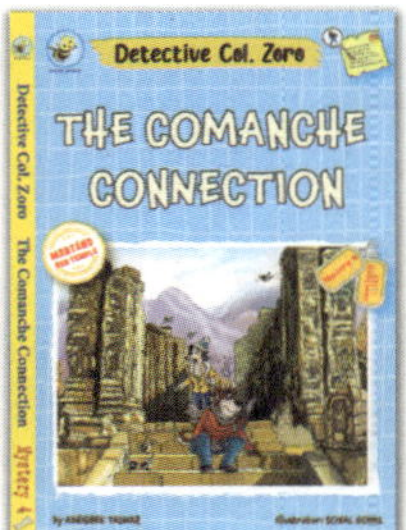

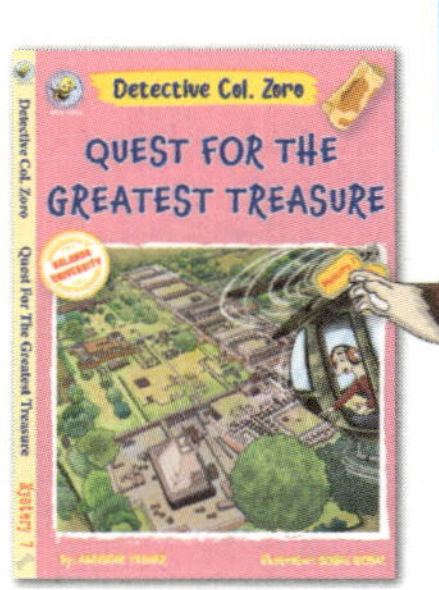

DRAW YOUR STORY

If you enjoyed reading about these adventures of Biplob the Bumblebee, then you will love what comes next! Read the paragraph below:

> It was another lovely rainy day at Biplob's garden.
>
> As Biplob reached the well, he found farmer Balram standing there with a frown on his face. "What are you thinking farmer Balram?" asked Biplob.
>
> "I'm just thinking, Biplob. It is raining so well this monsoon season. But my well has such a small mouth that most of the water just flows away and goes waste, instead of collecting in it. I'm afraid once the rains stop, we may again have a problem in watering the garden and the fields," said farmer Balram.

Now, draw the scene described in it, on the opposite page. Remember, there is nothing right or wrong – use your imagination, draw and colour the scene, as you see it!

Scan this code to upload an image of your drawing
and you could get featured on our instagram page!

About the Author

Abhishek Talwar (AIEMA) is a certified environmentalist and author. He has the gift of narrating difficult concepts with simplicity and fun. This ease of narrative combined with the values of love for the environment, loyalty and friendship that shine through each story make these books truly entertaining and enriching for children.

"Dedicated to my loving wife, Ritika, for always believing.
And to my curious children, for inspiring these tales."

-Abhishek Talwar

About the Illustrator

Sonal Goyal completed her masters in fine arts from College of Art, Delhi. Having been creative head at a leading publishing house, she is now an independent artist. Her love for drawing and books has led her to creating cute and lovable illustrations that spread smiles.

www.biplobworld.com